2023 DW: The Newly Discovered Asteroid That Might Hit The Earth

Table of Contents

Chapter 1

ASTEROID 2023 DW

The space rock 2023 DW was simply found in late February. Yet, NASA says it's following it closely to find out about its orbital way because the space rock "has a tiny possibility of affecting Earth" in 23 years.

The space rock's width is recorded at almost 50 meters — generally the size of an Olympic pool. It requires 271 days to circle the sun.

That's what NASA says after another article is first found, "it requires a little while of information to diminish the vulnerabilities and satisfactorily foresee their circles a very long time into what's to come."

Concerning how little the opportunity of effect is at present assessed to be, NASA puts it at "1 out of 560 chances of effect." Put another way, there is just a 0.18% possibility of hitting Earth or a 99.82% opportunity that the space rock will streak innocuously past our planet.

Space rock 2023 DW presently beat the "Hazard Rundown" kept up with by the European Space Organization — a program of 1,450 close Earth objects "for which a non-zero effect likelihood has been processed."

Comparatively measured objects have hit Earth

Regardless of whether the space rock was to strike our planet, making a comprehensively calamitous event wouldn't be normal. Objects of comparable size have hit Earth previously, including the effect around quite a while back that left the Meteor Cavity in cutting-edge Arizona.

Furthermore, in 1908, Siberian forestland was crushed in the "Tunguska Occasion" which destroyed 800 square miles of land and impacted 80 million trees, leaving them to spread in a spiral example.

The study of risk appraisals
There are two effect peril scales: the Palermo Scale, which experts use to confer a granular gander at the potential dangers presented by close earth objects, and the more seasoned Torino Scale,

which utilizations variety codes and a 0-10 rating to convey potential dangers to general society.

NASA makes sense of how the Palermo Scale functions:

"For comfort the scale is logarithmic, thus, for models, a Palermo Scale worth of - 2 shows that the recognized potential effect occasion is just 1% as possible as an irregular foundation occasion happening in the mediating years, a worth of zero demonstrates that the single occasion is similarly basically as compromising as the foundation peril, and a worth of +2 shows an occasion that is multiple times more probable than a foundation influence by an item to some degree as enormous before the date of the expected effect being referred to."

Space rock 2023 DW is one of just three items that at present have a Palermo Scale esteem more noteworthy than - 3, recorded at - 2.17 on the ESA's site.

Chapter 2

EFFECTS OF ASTEROIDS

Space rocks needn't bother with to be enormous to be damaging

Most meteors wreck on passage into Earth's air, the grating with the air disintegrates them or breaks them into more modest pieces that are by and large not perilous. In any case, like clockwork, plus or minus, a space rock the size of a football field, which is sufficiently large to cause huge neighborhood harm, enters our environment. Nonetheless, even little divine bodies that enter Earth's climate can cause outrageous harm assuming they detonate before arriving at the ground. In 1908, a meteor maybe 20 meters across exploded over the Siberian woods in Russia. The power of the impact, known as the Tunguska occasion, was comparable to that of a 12-megaton bomb and evened out an area of around 830 square miles. One more remembered to be of comparative size about quite a while back cleared out the city of Tall el-Hammam in the Jordan Valley and the encompassing region couldn't be cultivated for any less than 300 years. This occasion is credited with being the motivation for the scriptural story of Sodom and Gomorrah.

quite a while back today The Chelyabinsk meteor was a superbolide that placed Earth's air on 15 February 2013. It was momentarily more brilliant than the Sun, apparent up to 100 km (62 mi) away.

Strategies to safeguard Earth from space rocks

There are worldwide endeavors to shield Earth from possible dangers from space. NASA sent off the Twofold Space rock Divert mission, or DART, in 2021 fully intent on changing the circle of space rock. The organization's rocket was on a self-destruction mission to collide with the more modest moonlet of a parallel space rock framework, including Didymos and Dimorphos, around 7 million miles from Earth.

Chapter 3

MORE ON ASTEROID 2023DW

So these outcomes are incredibly starters, with not many perceptions.

Yet, all things considered, as of this second, space experts are assessing the space rock as 165 feet (50 meters) in distance across. (Watch a video of size examinations in space rocks.) While 50 meters is nothing close to a planet-finishing estimated rock, it would make an imprint in anything district it hit, if without a doubt it ought to strike.

Truth be told, the Chelyabinsk meteor - which detonated in the air over Russia in 2013 - was a little under a portion of the size of the gauge for 2023DW.

At present a 1 on the Torino scale
Likewise, 2023DW presently has a Torino score of 1. That is an extremely interesting score. However, even the extremely uncommon Torino score of 1 (rather than 0) is not something to be stressed over. A score of 1 method:

A standard disclosure wherein a pass close to Earth is anticipated that represents no uncommon degree of risk. Current computations show an impact is incredibly far-fetched with no reason for public consideration or public concern. New adaptive perceptions probably will prompt reassignment to Even out 0.

Richard Binzel of the Massachusetts Establishment of Innovation, the innovator of the Torino scale, talked with EarthSky.

He made sense of that it's typical to see the numbers go up as we look further into the article's circle. In these cases, we are simply restricting the line of vulnerability. The nearer the space rock will be to Earth, the more the numbers will go up until we dispense with its chance of hitting Earth, at which case the space rock will drop down to classification zero.

This has been the situation with 2023DW. It initially had around a 1-in-700 possibility of hitting Earth, which is presently 1-in-560 as we get more familiar with its circle. Concerning when we'll get the normal "all reasonable" that 2023DW will not hit Earth, Binzel said:

Hopefully, we'll know more in half a month.

Chapter 4

HOW TO SURVIVE A KILLER ASTEROID IMPACT

NASA affirms it can divert a lethal space rock from hitting Earth - however, this is the very thing that YOU ought to do if the office's main goal falls flat

NASA affirmed Monday it can redirect a destructive space rock going to Earth Assuming that the office falls flat, four endurance strategies could assist you with remaining alive

Review affirms NASA's space rock diversion test was a triumph

While NASA as of late affirmed it can redirect an incredible space rock off a way toward Earth, it leaves individuals considering how they would make due if one hit our planet.

The last devastating effect happened quite a while back, killed the dinosaurs, and a few researchers accept we are expected for another 'huge one.'

With this inescapable destruction approaching mainstream researchers, specialists are energetically chipping away at how-to advisers to assist mankind with remaining alive.

The initial step is to obliterate the space rock before it is past the point of no return, and keeping in mind that the American space organization appears to take care of this, more than 2,000 possibly unsafe space rocks are holding on to sneak by its radar.

NASA sent off its Twofold Space rock Redirection Test (DART) in 2022 for mankind's most memorable planetary safeguard mission, named NASA's 'Armageddon second.'

The art's objective was a moonlet called Dimorphos orbiting its parent space rock, Didymos.

On September 26, the world looked as DART took off 15,000 miles each hour toward Dimorphos to push it off its circle.

Furthermore, on Walk 1, 2023, NASA affirmed the mission was a crushing achievement.

The space organization's cooler estimated satellite figured out how to shave 33 minutes off the circle of a 520 all-inclusive space rock - almost multiple times more noteworthy than what was anticipated.

Researchers from the Northern Arizona College said: 'This fills in as a proof-of-idea for the motor impactor procedure of planetary protection, DART expected to show the way that a space rock could be designated during a high-velocity experience and that the objective's circle could be changed.'

The likelihood of a space rock the size of the dinosaur-killing Chicxulub hitting our planet is one each 100 to 200 million years - yet the occasion isn't unimaginable.

If NASA neglects to avoid the giant space rock, specialists expressed the following best is to leave the effect zone and get away from waterfront regions.

NASA sent off its Twofold Space rock Redirection Test (DART) in 2022 for humankind's most memorable planetary safeguard mission, named NASA's 'Armageddon second.' The specialty's objective was a moonlet called Dimorphos (envisioned) circumnavigating its parent space rock, Didymos

NASA affirmed Monday that it effectively redirected a space rock in space, however, individuals actually can't help thinking about how they might make due assuming the office's central goal comes up short.

Imagined is the second the art crushed into the space rock in NASA's DART mission

One tip is to get away from seaside regions because of the transcending waves that the effect would set off. A 9.1-greatness undersea seismic tremor struck off Japan's coast on Walk 11, 2011 (envisioned)

Since Earth is 71% water, there is a more noteworthy possibility the space rock would fall into the sea.

Also, when it does, the effect would make transcending torrents that would immerse all close by land.

A 9.1-size undersea seismic tremor struck off Japan's coast on Walk 11, 2011.

It caused a strong tidal wave that caused a complete implosion of 3 reactors at the Fukushima Daiichi atomic plant and a huge number of occupants had to empty the region.

Official figures delivered in 2021 announced 19,747 passings, 6,242 harmed, and 2,556 individuals missing from this tidal wave fiasco

Also, those made from a space rock effect would much pulverize.

One more way to stay alive is by looking for underground asylum.

In any case, when a space rock hits, it discharges residue, flotsam, jetsam, and, surprisingly, poisonous gases that would wait in the environment for a long time - even many years.

The dugouts are underground, with simply a doorway on a superficial level

Assuming you have made it this far, following effect, researchers recommend remaining in the asylum until you can demonstrate the climate is protected

Also, researchers accept the most secure spot would be an underground fortification.

Dugouts can be costly, going from $20,000 to 1,000,000 bucks and upwards, making these safe houses even more of an extravagance as opposed to a need.

A complex of these offices in South Dakota houses 10,000 individuals and is being hailed as the 'plan B for humankind.'

Assuming you have made it this far, following effect, researchers recommend remaining in the sanctuary until you can demonstrate the climate is protected.

This should be possible by continually looking at the air outside to guarantee ordinary levels, guaranteeing flames and floods have died down and the corrosive downpour isn't tumbling from the sky.

One choice disposes of the requirement for a survival manual and pulls motivation from the Netflix film Don't Turn Upward - holding on until it is past the point of no return and simply tolerating the finish of humankind

Delivered in January 2022, the film highlights Leonardo DiCaprio and Jennifer Lawrence, depicting two space experts attempting to beat the clock to caution the world about an approaching space rock to obliterate the planet.

They distinguish a comet coming for planet Earth in a half year and 14 days and endeavor to caution the universe of the disclosure, yet individuals couldn't care less about terrible news about the future - and Earth and all living things are obliterated in a wad of flares.